FUN FACT FILE:
INVENTIONS THAT CHANGED THE WORLD

20 FUN FACTS ABOUT THE STEAMBOAT

BY JANEY LEVY

Gareth Stevens
PUBLISHING

Please visit our website, www.garethstevens.com. For a free color catalog of all our high-quality books, call toll free 1-800-542-2595 or fax 1-877-542-2596.

Library of Congress Cataloging-in-Publication Data
Names: Levy, Janey, author.
Title: 20 fun facts about the steamboat / Janey Levy.
Other titles: Twenty fun facts about the steamboat
Description: Buffalo, New York : Gareth Stevens Publishing, [2024] | Series: Fun fact file: inventions that changed the world | Includes index.
Identifiers: LCCN 2022053141 (print) | LCCN 2022053142 (ebook) | ISBN 9781538287743 (library binding) | ISBN 9781538287736 (paperback) | ISBN 9781538287750 (ebook)
Subjects: LCSH: Steamboats–History–Juvenile literature.
Classification: LCC VM615 .L48 2024 (print) | LCC VM615 (ebook) | DDC 623.82/436–dc23/eng/20221102
LC record available at https://lccn.loc.gov/2022053141
LC ebook record available at https://lccn.loc.gov/2022053142

Published in 2024 by
Gareth Stevens Publishing
2544 Clinton Street
Buffalo, NY 14224

Designer: Leslie Taylor
Editor: Therese Shea

Photo credits: Cover (main), pp. 1 & 5 pisaphotography/Shutterstock.com; Series Art (file folder) David Smart/Shutterstock.com, (binder clip) luckyraccoon/Shutterstock.com, (wood grain background) ARENA Creative/Shutterstock.com; p. 5 (horseback) w:en:User:PHG/https://en.wikipedia.org/wiki/File:PazyrikHorseman.JPG, (walking) Elzbieta Sekowska/Shutterstock.com, (boats) Estormiz/https://commons.wikimedia.org/wiki/File:Dugout_boats_Kierikki_Centre_Oulu_20130526.JPG, (carriage) chippix/Shutterstock.com; p. 6 (inset) https://commons.wikimedia.org/wiki/File:Aeolipile_illustration.png, (main) Love Employee/Shutterstock.com; p. 7 https://ethw.org/File:Newcomen_Figuier.jpg; p. 8 (illustration) J J Osuna Caballero/Shutterstock.com, (Watt) Antonia Reeve/https://en.wikipedia.org/wiki/File:James-watt-1736-1819-engineer-inventor-of-the-stea.jpg; p. 9 (Trevithick's) chris55/https://en.wikipedia.org/wiki/File:TrevithicksEngine.jpg, (steamer) https://commons.wikimedia.org/wiki/File:Driving_to_top_of_Mt_Washington_1899.jpg; (steam tractor) Acroterion/https://commons.wikimedia.org/wiki/File:Frick_Eclipse_steam_tractor_VA1.jpg; (Vapeur) Joe deSousa/https://commons.wikimedia.org/wiki/File:Joseph_Cugnot%27s_1770_Fardier_à_Vapeur,_Musée_des_arts_et_métiers,_Paris_2015.jpg, (airship) Mike Young/https://en.wikipedia.org/wiki/File:GiffardAirship.JPG; p. 10 Nastasic/istockphoto.com; p. 11 World Imaging/https://en.wikipedia.org/wiki/File:D%27AbbansSteamshipModel.jpg; p. 12 Smithsonian American Art Museum, Bequest of Eugene A. Rumsey and Brothers/library.si.edu; p. 13 Dr. Karl-Heinz Hochhaus/https://commons.wikimedia.org/wiki/File:2013_11_22_Versuchsboot_J_Fitch_1785-IMG_2729_P_K_S.JPG; p. 14 (both) Everett Collection/Shutterstock.com; p. 16 http://steamboattimes.com/artwork_2.html/https://commons.wikimedia.org/wiki/File:Hippolyte_Sebron_-_Bateaux_A_Vapeur_Géants_1853.jpg; p. 17 (map) Peter Hermes Furian/shutterstock.com; p. 18 (stamp) neftali/Shutterstock.com, (Stevens) John Stevens by an unidentified artist, c. 1830, oil on canvas, from the National Portrait Gallery - NPG-NPG 75 13Stevens d1.jpg; p. 19 neftali/Shutterstock.com; p. 20 George Atkinson Jnr/https://commons.wikimedia.org/wiki/File:SS_Sirius_(1837).jpg; p. 21 (sidewheeler) elmm/Shutterstock.com, (sternwheeler) EB Adventure Photography/Shutterstock.com; p. 22 Henry Anthony/https://en.wikipedia.org/wiki/File:PS_Bristol_saloon.jpg; p. 23 (Sultana) Thomas W. Bankes/ttps://commons.wikimedia.org/wiki/File:Civil_War_Steamer_Sultana_tintype,_1865.png, (illustration) https://commons.wikimedia.org/wiki/File:Sultana_Disaster.jpg; p. 24 Internet Archive Book Images/https://commons.wikimedia.org/wiki/File:Life_and_liberty_in_America;_(1859)_(14776963692).jpg; p. 25 http://collections.rmg.co.uk/File:H.M. Steam sloops Rattler and Alecto towing stern to stern, for the purpose of testing the relative powers of the Screw Propeller and the Paddle Wheel - in the North Sea during a perfect calm on the 3rd of April 1845, RMG PY0923.jpg; p. 26 (French) https://en.wikipedia.org/wiki/File:Callie-Leach-French-ca-1890s-1.jpg, (steamboat) https://commons.wikimedia.org/wiki/File:ShowboatFrenchsNewSensation.jpg; p. 27 Andrew J. Russell/https://commons.wikimedia.org/wiki/File:East_west_shaking_hands_by_russell.jpg; p. 29 U.S. Navy Photo by John Narewski/https://en.wikipedia.org/wiki/File:USS Oregon (SSN-793) pulls into Submarine Base New London in Groton, Connecticut (USA), on 1 March 2022 (220301-N-UM744-002).JPG.

Printed in the United States of America

CPSIA compliance information: Batch #CSGS24: For further information contact Gareth Stevens at 1-800-542-2595.

CONTENTS

Words in the glossary appear in **bold** type the first time they are used in the text.

SLOW MOTION

How do you get around? Maybe you use a bike, car, bus, train, or plane. These forms of **transportation** carry you quickly from place to place. But for centuries, transportation was slow.

Long ago, people walked, rode animals, or used simple wheeled carts. Human power or wind moved boats. The steamboat offered a major advance in **vehicles**. But it was possible only because of the invention of the steam engine. Inside this book, you'll discover fun facts about steam engines and steamboats!

EARLY DAYS OF TRAVEL

animals

walking

simple boats

carriage

The very first humans walked to get around. About 10,000 years ago, people created the earliest boats. They started using animals about 6,000 years ago. Then, they invented the wheel about 5,500 years ago!

STEAM POWER!

FUN FACT: 1

THE VERY FIRST STEAM ENGINES POWERED TOYS.

Almost 2,000 years ago, a Greek scientist named Hero invented the first steam engine. But ancient Greeks used steam to create movement in small animal or human figures, not to power transportation.

This is a modern copy of Hero's steam engine. When water in the base was heated, steam rose through the two pipes to the ball. The steam shot out the two bent tubes on the ball. This caused the ball to spin.

Thomas Newcomen invented the next improvement in the steam engine in 1711. This was needed because Savery's engines sometimes caused explosions!

FUN FACT: 2

THE FIRST USEFUL STEAM ENGINES WERE INVENTED TO MOVE WATER.

Spanish inventor Jerónimo de Ayanz created a steam engine to remove water from flooded **mines** in 1606. British inventor Thomas Savery **patented** a more famous steam engine in 1698 to do the same thing.

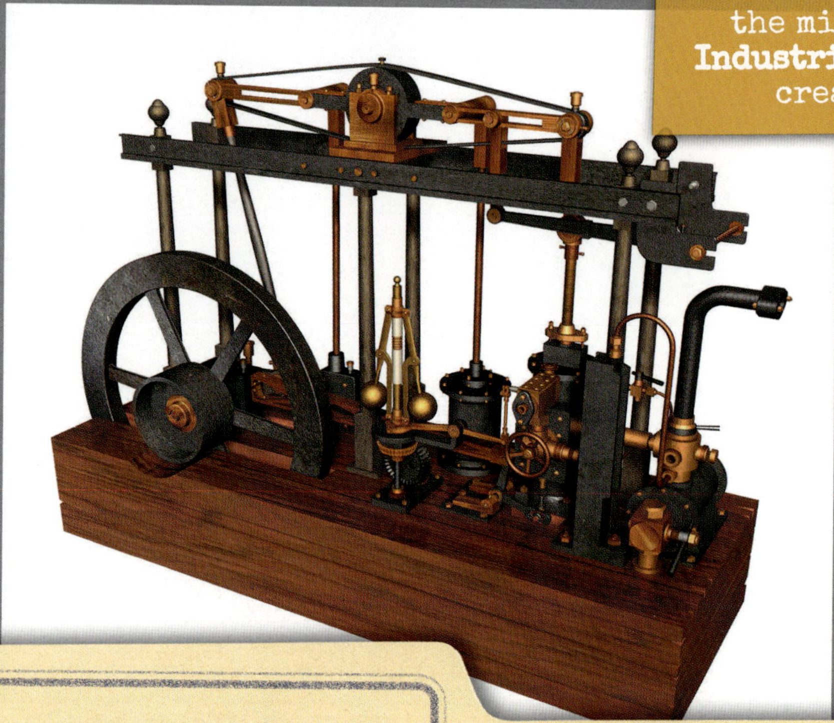

Watt's steam engine powered the mills and factories of the **Industrial Revolution** and helped create the modern world.

James Watt

FUN FACT: 3

THE WORD "HORSEPOWER"—WHICH TELLS HOW MUCH POWER A CAR HAS—WAS FIRST USED TO DESCRIBE A STEAM ENGINE.

Scottish inventor James Watt created an engine that wasted less steam than Thomas Newcomen's. He sold it in 1776 by bragging about how many horses it could replace.

EARLY STEAM-POWERED VEHICLES

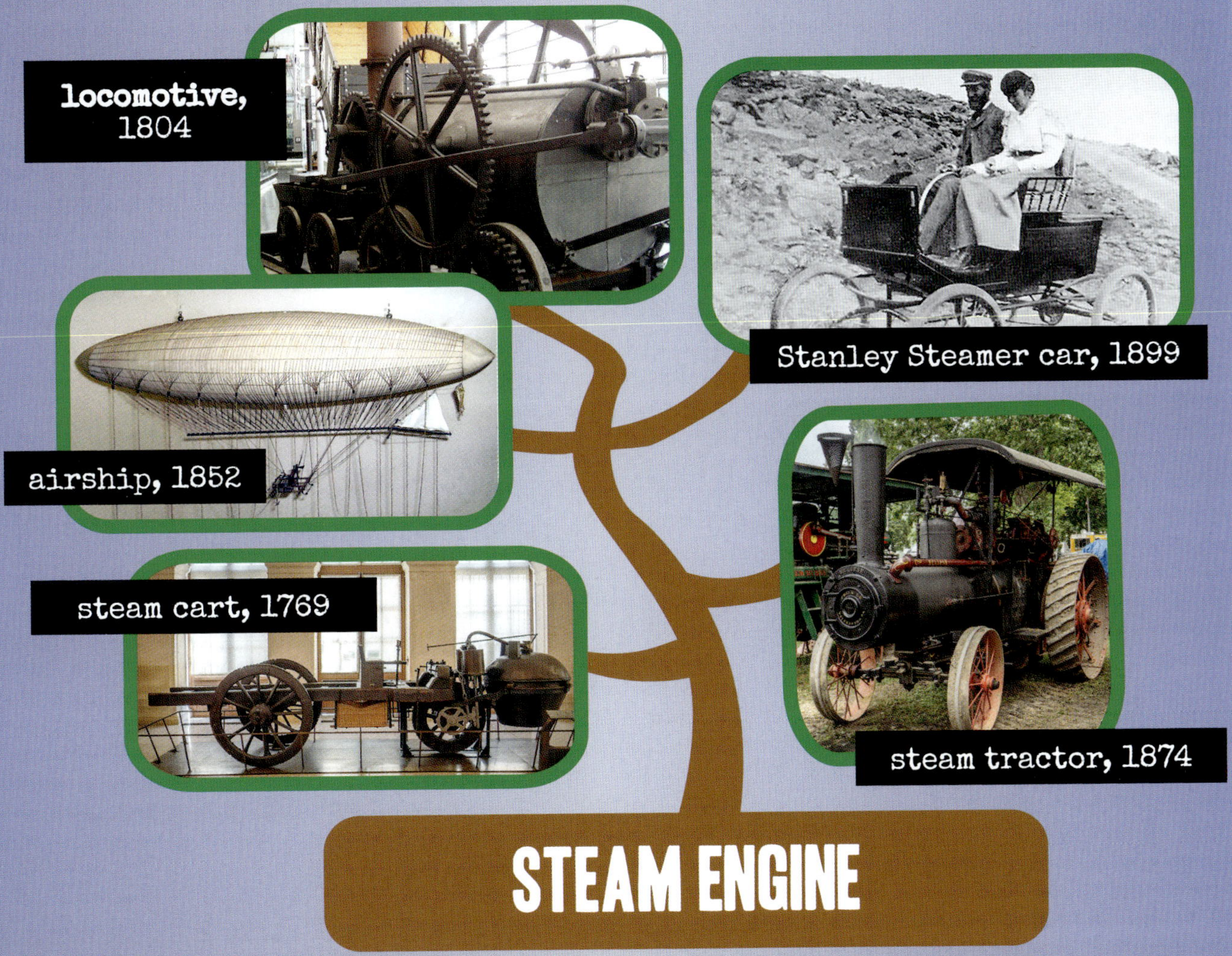

Inventors soon began to use steam engines to power vehicles. These are a few early examples.

INVENTING THE STEAMBOAT

FUN FACT: 4

THE IDEA FOR A STEAMBOAT WAS FIRST SUGGESTED IN 1543—OVER 60 YEARS BEFORE THE FIRST USEFUL STEAM ENGINE WAS INVENTED!

In 1707, inventors began real efforts to create a steamboat. But no one had success until after James Watt patented his improved steam engine in 1765.

Watt worked with **engineer** Matthew Boulton to produce his steam engine in 1776. They continued to improve it and created one like this engine.

This is a model of Jouffroy d'Abbans's steamboat.

FUN FACT: 5

THE INVENTOR OF THE FIRST SUCCESSFUL STEAMBOAT HAS BEEN MOSTLY FORGOTTEN.

French engineer Claude-François-Dorothée, marquis de Jouffroy d'Abbans, invented a steamboat he ran on a French river in 1783. He couldn't get a patent for it, so he gave up working on it.

This painting of Rumsey was done around 1790. The American engineer died about two years later while trying to get people to finance his invention. Doctors said his death was caused by overworking his brain!

FUN FACT: 6

JET PROPULSION MOVED AN EARLY AMERICAN STEAMBOAT!

In 1787, James Rumsey's steamboat used a steam engine to operate a pump. The pump pulled in water from below the boat and shot it out the back. This created force that pushed the boat forward.

FUN FACT: 7

A STEAMBOAT USING PADDLES SET THE PATTERN FOR LATER STEAMBOATS.

Fitch soon built more steamboats to carry people and goods between Philadelphia, Pennsylvania, and Burlington, New Jersey. But his boats were so costly to build and operate that they weren't successful.

The same year Rumsey displayed his steamboat, American inventor John Fitch presented one moved by paddles. But unlike Fitch's boat, later steamboat paddles would be fixed to wheels—paddle wheels.

FUN FACT: 8

ROBERT FULTON IS CALLED THE FATHER OF THE STEAMBOAT EVEN THOUGH HE DIDN'T INVENT IT.

The *Clermont* traveled up the river at a then-quick speed of 5 miles (8 km) per hour. It was faster than the 3 miles (4.8 km) per hour that an adult walks.

Fulton made steamboats a successful business. His boat, the *Clermont*, first traveled up the Hudson River—from New York City to Albany, New York—in 1807. Soon people were paying to make the journey!

INVENTING THE STEAMBOAT

1543	The first idea for a steamboat is recorded.
1776	James Watt sells a better steam engine.
1783	Claude-François-Dorothée, marquis de Jouffroy d'Abbans, invents a steamboat he runs on a French river.
1787	James Rumsey invents a jet-propulsion steamboat. John Fitch invents a steamboat moved by paddles.
1789	Scottish inventor William Symington creates a steamboat.
1792	Elijah Ormsbee and David Wilkinson build a steamboat that runs on Narragansett Bay in Rhode Island.
1793	Samuel Morey runs a paddle-wheel steamboat from Hartford, Connecticut, to New York City.
1802	Symington builds a paddle-wheel steamboat, the *Charlotte Dundas*.
1804	John Stevens invents a steamboat.
1807	Robert Fulton's *Clermont* makes its first trip up the Hudson River.

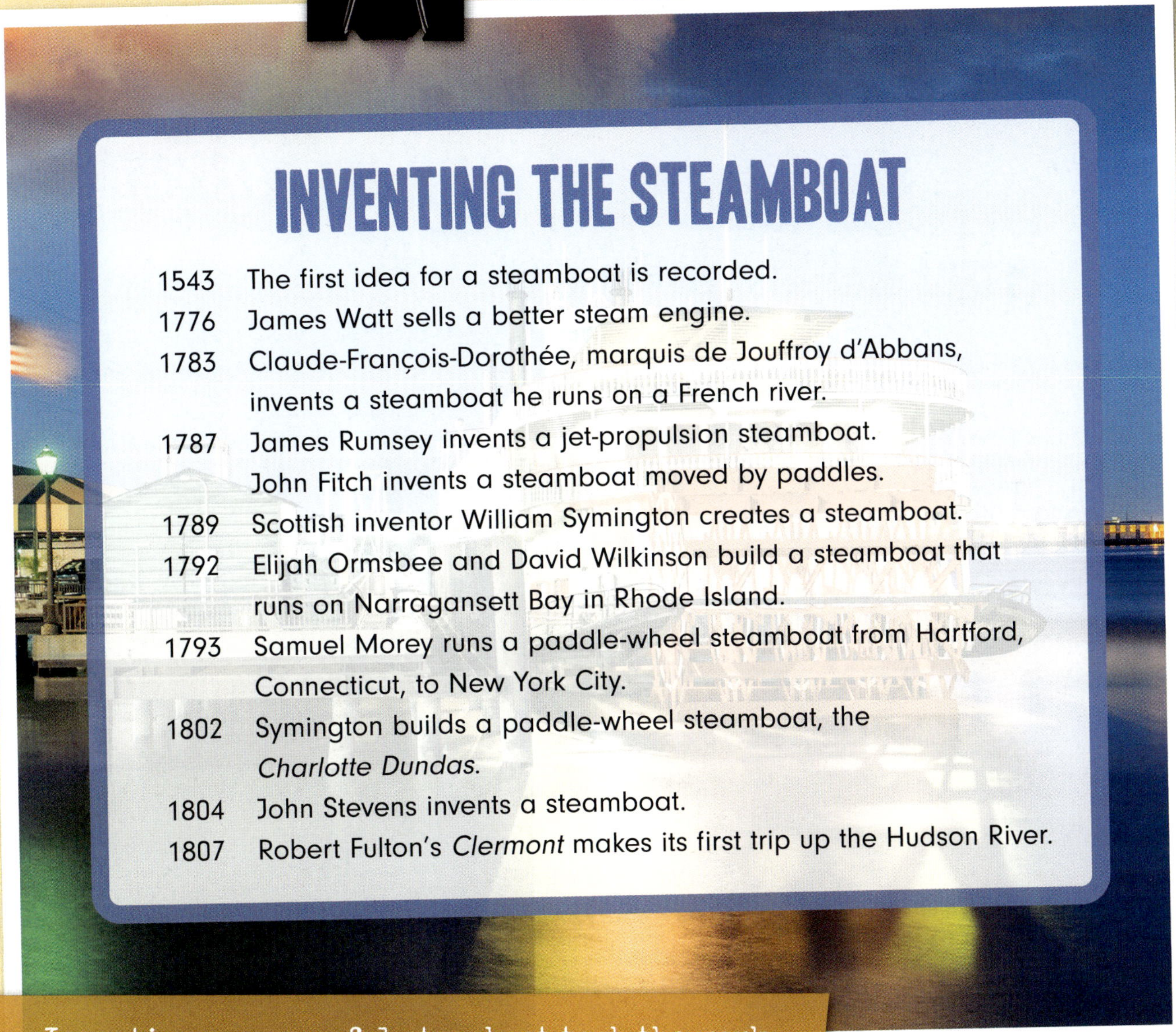

Inventing a successful steamboat took the work of a lot of people over many years.

STEAMBOATS CONQUER THE WATERS

FUN FACT: 9

THE PRICE OF TRAVELING ONE WAY ON THE *CLERMONT* IN 1807 WAS $7—ABOUT $180 TODAY!

The *Clermont* started carrying people and goods regularly between New York and Albany on September 4, 1807. Interest in steamboats exploded after its success.

The number of steamboats arriving in New Orleans, Louisiana, rose from 20 in 1814 to 1,200 in 1834.

By 1812, Fulton was running steamboats on the Hudson River and the Mississippi River, plus four more rivers and the Chesapeake Bay.

FUN FACT: 10

STEAMBOATS FROM DIFFERENT COMPANIES CRASHED INTO EACH OTHER ON THE HUDSON RIVER BECAUSE THEY "PLAYED CHICKEN."

Only Fulton had the right to operate steamboats on the Hudson River. But other steamboat companies soon appeared, crowding the river. People knew they could make money from the invention.

FUN FACT: 11

THE FIRST STEAMBOAT JOURNEY ON THE OCEAN HAPPENED IN 1809.

Inventor John Stevens's boat, the *Phoenix*, sailed on the Atlantic Ocean but didn't cross it. It stayed near the coast, traveling from Hoboken, New Jersey, to Philadelphia, Pennsylvania.

John Stevens

The *Phoenix* was the first financially successful steamboat built completely in the United States. The *Clermont* had used a steam engine built in England.

With the *Washington*, Shreve created the form used for later Mississippi River steamboats. The steam engine was placed high above the water line, the boat's bottom was flat, and a tall second deck was added.

FUN FACT: 12

IN 1816, A "FAST" STEAMBOAT TRIP FROM NEW ORLEANS, LOUISIANA, TO LOUISVILLE, KENTUCKY, TOOK 25 DAYS.

Henry Miller Shreve's steamboat *Washington* accomplished this feat. Before steamboats, boats took about 120 days to travel the same distance! But by 1853, the trip took only 4.5 days.

FUN FACT: 13

THE FIRST STEAMBOAT TO CROSS THE ATLANTIC OCEAN HAD TO USE SAILS FOR PART OF THE TRIP.

Shown here is the *Sirius*, the first steamboat to cross the Atlantic Ocean using only steam power in 1838.

In 1819, the *Savannah* traveled from the U.S. state of Georgia to St. Petersburg, Russia. But it ran out of the coal it needed to burn to produce steam.

sidewheeler

sternwheeler

Steamboats came in two basic types based on where their paddle wheels were placed. Sidewheelers had paddle wheels on each side. Sternwheelers had a paddle wheel at the stern, or rear.

FUN FACT: 14

DIFFERENT KINDS OF STEAMBOATS WERE BUILT TO DO DIFFERENT JOBS, INCLUDING ENTERTAIN!

Packet boats carried people, goods, and mail. Snagboats removed snags, or dangerous underwater objects, that could harm boats. Showboats brought music and plays to river towns.

FUN FACT: 15

LARGE STEAMBOATS COULD BE AS FANCY AS THE HOUSE OF A RICH PERSON.

Large steamboats had big public rooms decorated with beautiful rugs, paintings, and glass lights. Famous cooks prepared food, and maids served the guests. Boats hired musicians to play for travelers too.

The steamboat shown here, the *Bristol*, was so large it had room for 1,000 travelers!

DANGER!

FUN FACT: 16

SOME STEAMBOATS MIGHT HAVE BEEN FANCY, BUT THIS KIND OF TRAVEL WAS DANGEROUS!

Three **boilers** on the steamboat *Sultana* exploded on the Mississippi in 1865, killing as many as 1,800 people. It was the deadliest boat accident in U.S. history.

The boilers where the steam was produced sometimes exploded. Native American attacks occurred in some places. Snags sunk boats. And foolish steamboat captains created more danger by racing each other!

FUN FACT: 17

THE AVERAGE STEAMBOAT LASTED LESS THAN 5 YEARS.

Between 1830 and 1839, 272 steamboats that were less than three years old were destroyed. The steamboat *Princess* was only four years old when all four boilers exploded in 1859.

The dangers steamboats faced put an end to many of them. Steamboats also had short lives because they weren't built well to begin with. And they often weren't well cared for.

FUN FACT: 18

A TUG-OF-WAR CHANGED OCEAN STEAMBOATS!

The back ends of the steamboats were tied together. The *Rattler* won by pulling the *Alecto* at nearly 3.5 miles (5.6 km) per hour—even though the *Alecto* was pulling in the opposite direction!

Engineers questioned if paddle wheels were the best way to move ocean steamboats. A tug-of-war was set up between two steamboats in 1845. The *Alecto* had paddle wheels. The *Rattler* used a **propeller**—and won easily!

FUN FACT: 19

AT LEAST FOUR WOMEN WERE STEAMBOAT CAPTAINS IN THE LATE 1800s.

This was a time when women had few job opportunities. Women also served as pilots, in charge of directing and guiding the boats. Callie French served as captain, pilot, and cook!

Steamboat captain Callie French, above, owned the steamboat *French's New Sensation*, shown at left.

The first railroad reaching from one side of the United States to the other was completed in 1869.

FUN FACT: 20

STEAM ENGINES MADE STEAMBOATS POSSIBLE *AND* ENDED THE AGE OF STEAMBOATS.

Experiments with steam locomotives began with Richard Trevithick in 1803. By the 1870s, trains with steam locomotives had taken over the movement of people and goods from steamboats.

MODERN STEAM POWER

Are you thinking that boats moved by steam belong to the past? Some of the U.S. Navy's most modern ships use steam power!

Nuclear-powered ships include submarines, or underwater ships. They use power produced by breaking apart the nucleus, or center, of atoms. This produces heat. The heat causes water to produce steam, which turns **turbines**, which then turn the ship's propeller. And that makes the ship move. So, steam power is alive in the age of nuclear energy!

The navy nuclear submarine the USS *Oregon* was commissioned, or officially made active, in 2022. Scientists are working on nuclear ships to carry travelers and goods.

GLOSSARY

boiler: A large container, such as a tub or pot, heated to produce steam in an engine.

engineer: Someone who plans and builds machines.

entertain: To perform or provide amusement for others.

Industrial Revolution: A period of social and economic change marked by advances in technology and science.

jet propulsion: The forward movement of a body by means of a jet, or burst of liquid or gas, from the rear.

locomotive: A vehicle that produces power to pull a line of other vehicles, such as a train.

mine: A pit or tunnel from which rocks or other matter is taken.

nuclear: Having to do with the power created by splitting atoms, or small bits of matter.

patent: To obtain an official paper that gives a person the rights to a design, machine, or process for a time. Also, the name for this paper.

play chicken: A dangerous game in which two vehicles drive toward each other to see which driver will move away first.

propeller: Paddle-like parts on a ship that spin in the water to move the ship forward.

transportation: A means of moving people or goods from place to place.

turbine: A motor operated by the movement of water, steam, or air.

vehicle: An object used for carrying people or goods.

FOR MORE INFORMATION

BOOKS

Barr, Catherine, and Steve Williams. *The Story of Inventions: A First Book About World-Changing Discoveries.* Minneapolis, MN: Frances Lincoln Children's Books, 2020.

Braun, Eric. *Can You Survive the* Sultana *Disaster? An Interactive History Adventure.* North Mankato, MN: Capstone Press, 2023.

Macaulay, David. *Crossing on Time: Steam Engines, Fast Ships, and a Journey to the New World.* New York, NY: Roaring Brook Press, 2019.

WEBSITES

John Fitch Was Granted a U.S. Patent for the Steamboat
www.americaslibrary.gov/jb/nation/jb_nation_fitch_1.html
Discover more about the history of steamboats.

Robert Fulton
www.ducksters.com/history/us_1800s/robert_fulton.php
Find out about the life of Robert Fulton, who developed the first commercially successful steamboat.

What Is a Steamboat?
wonderopolis.org/wonder/what-is-a-steamboat
Learn about steamboats, including how they worked and transformed the United States.

INDEX